Halloween

See, Think and Wonder

Deanna Pecaski McLennan, Ph.D.

For Cadence, Caleb and Quinn
I've loved all our Halloweens together!

Copyright © by Deanna Pecaski McLennan
First edition 2021

All rights reserved.

No part of this publication may be reproduced in any form, or by any means, electronic or mechanical, including photocopying, recording, or any information browsing, storage or retrieval system, without permission in writing from the author.

www.mrsmclennan.blogspot.ca

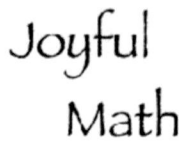

Joyful Math

Halloween is so much fun! I love exploring the autumn world around me. I discover math everywhere.

I **see** many cornstalks!

I **think** they are taller than me.

I **wonder** why they each grow a pattern.

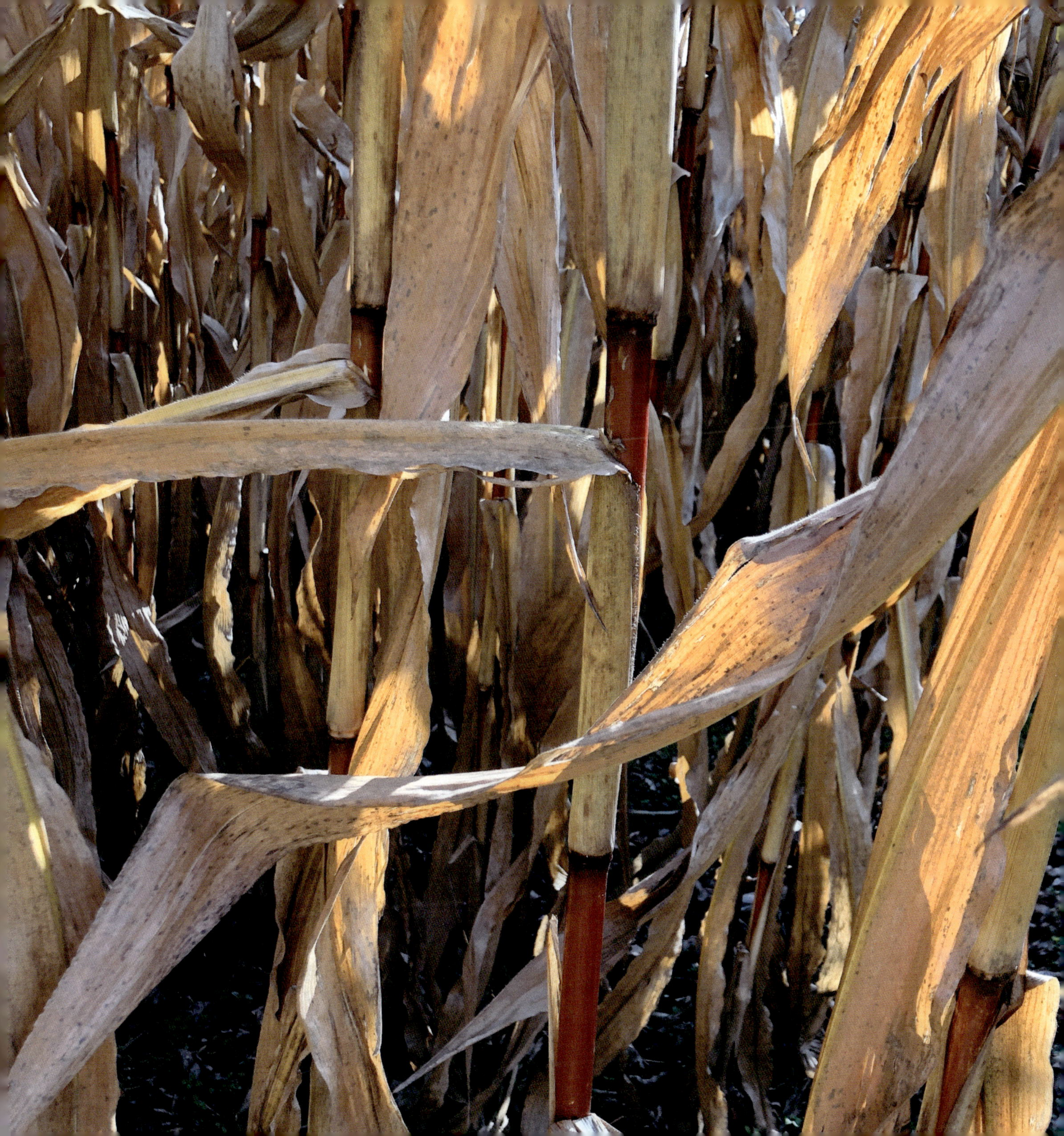

I **see** kernels of corn!

I **think** they grow in straight rows.

I **wonder** how many there are on each ear.

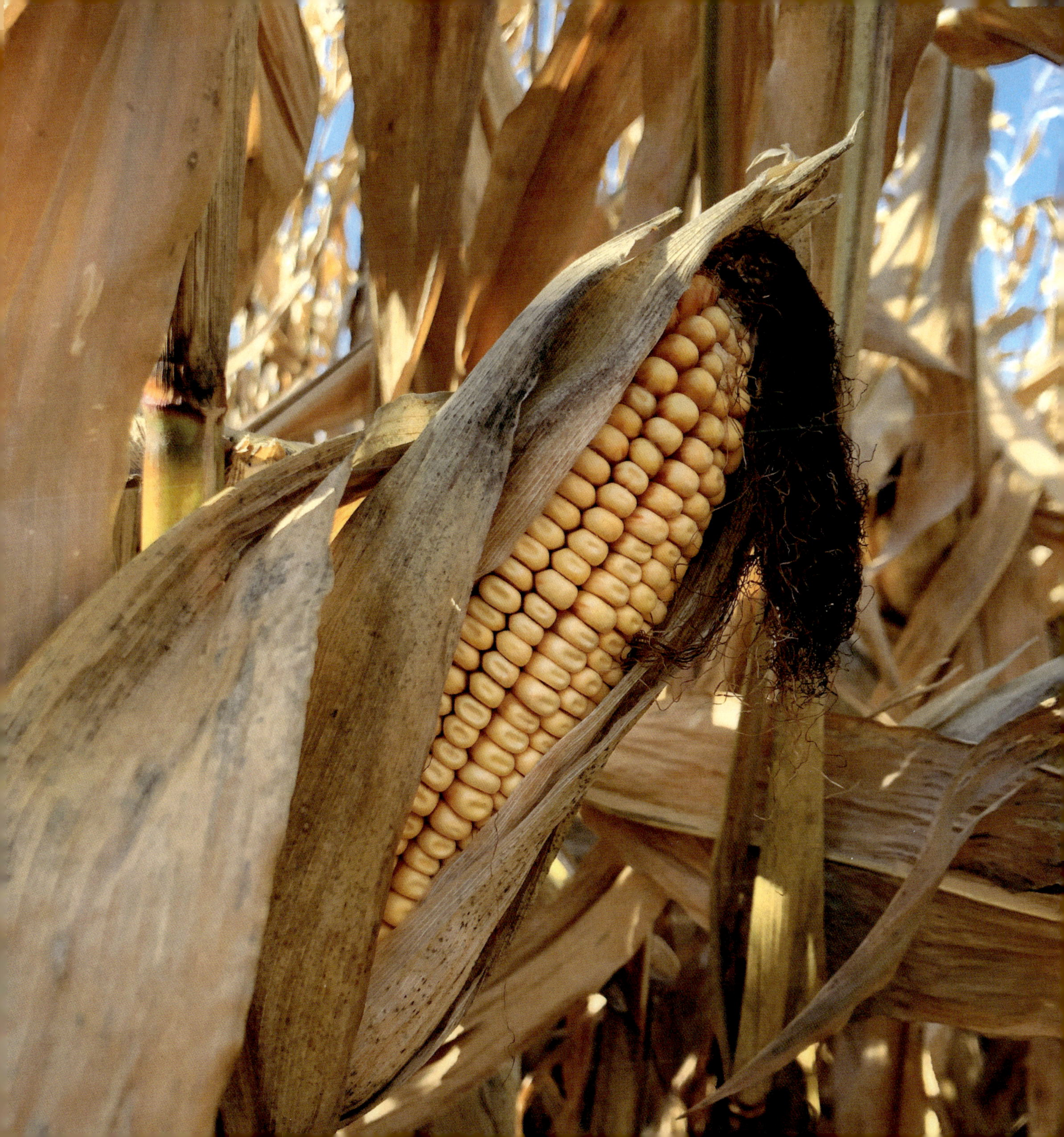

I **see** bales of hay!

I **think** they are rectangular prisms.

I **wonder** how they fit together in the stack.

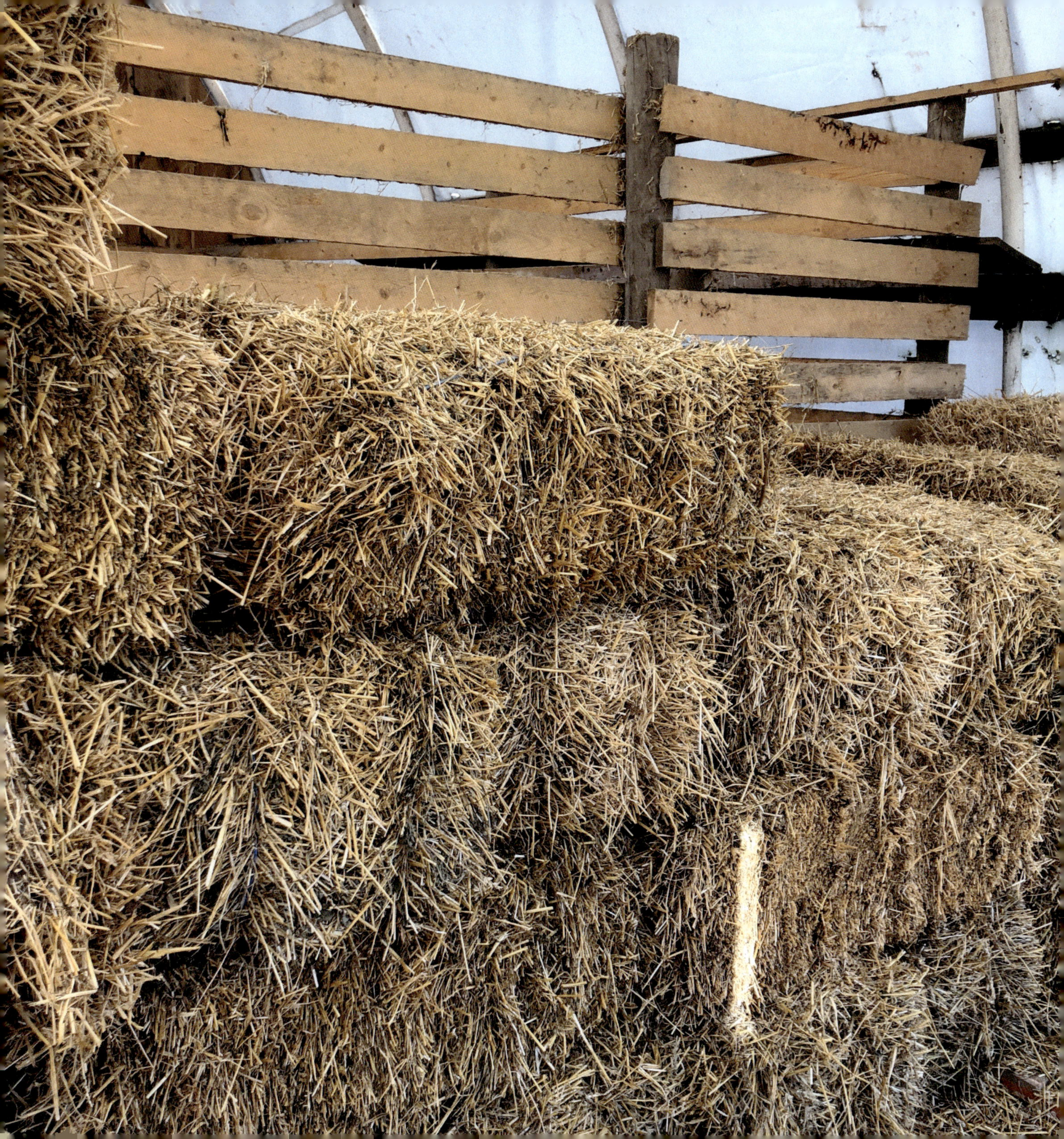

I **see** a big, black hog!

I **think** he is eating a whole pumpkin.

I **wonder** how many in total he eats each day.

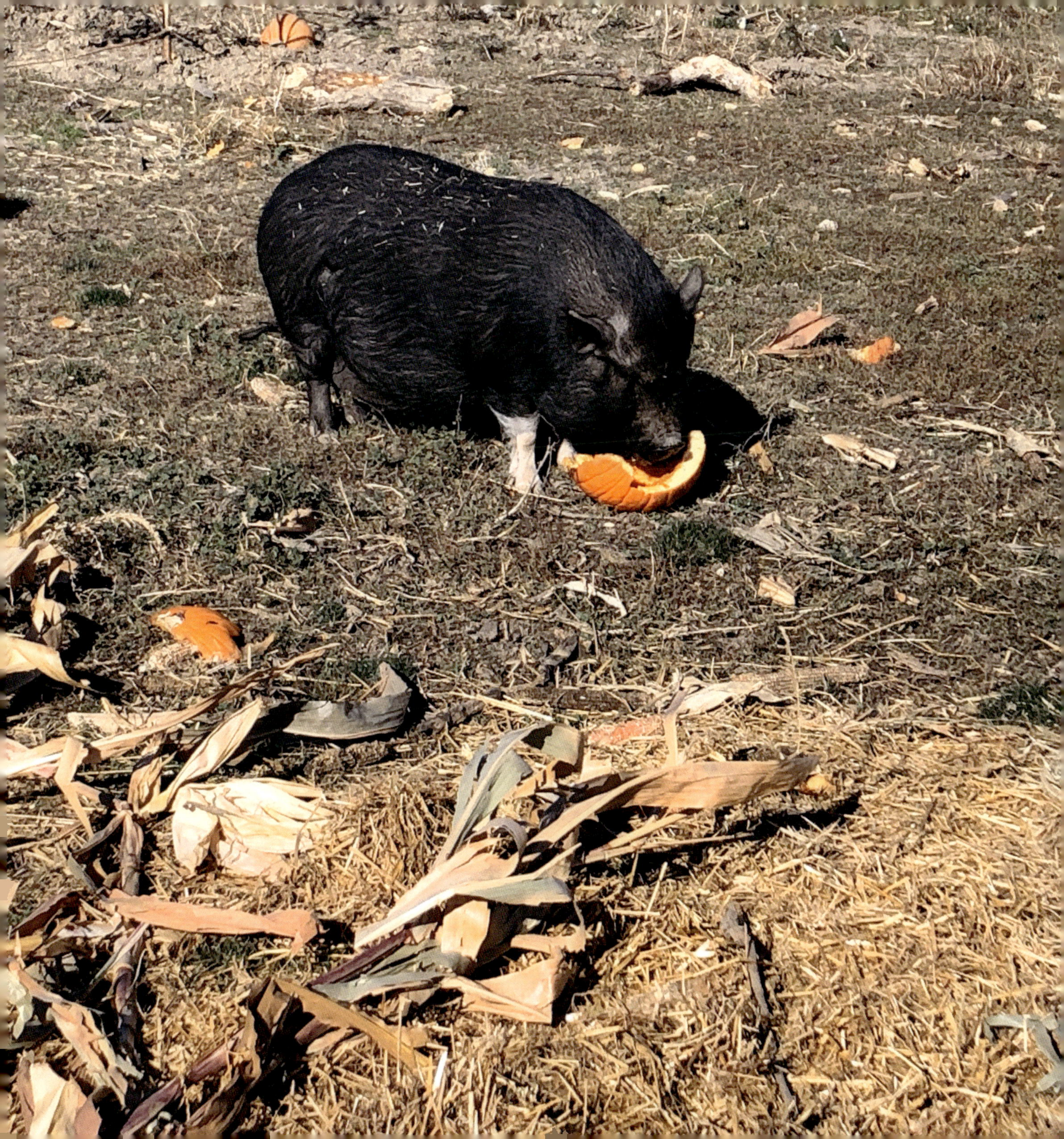

I **see** pumpkins in the field!

I **think** they are all different sizes.

I **wonder** which one is the heaviest of all.

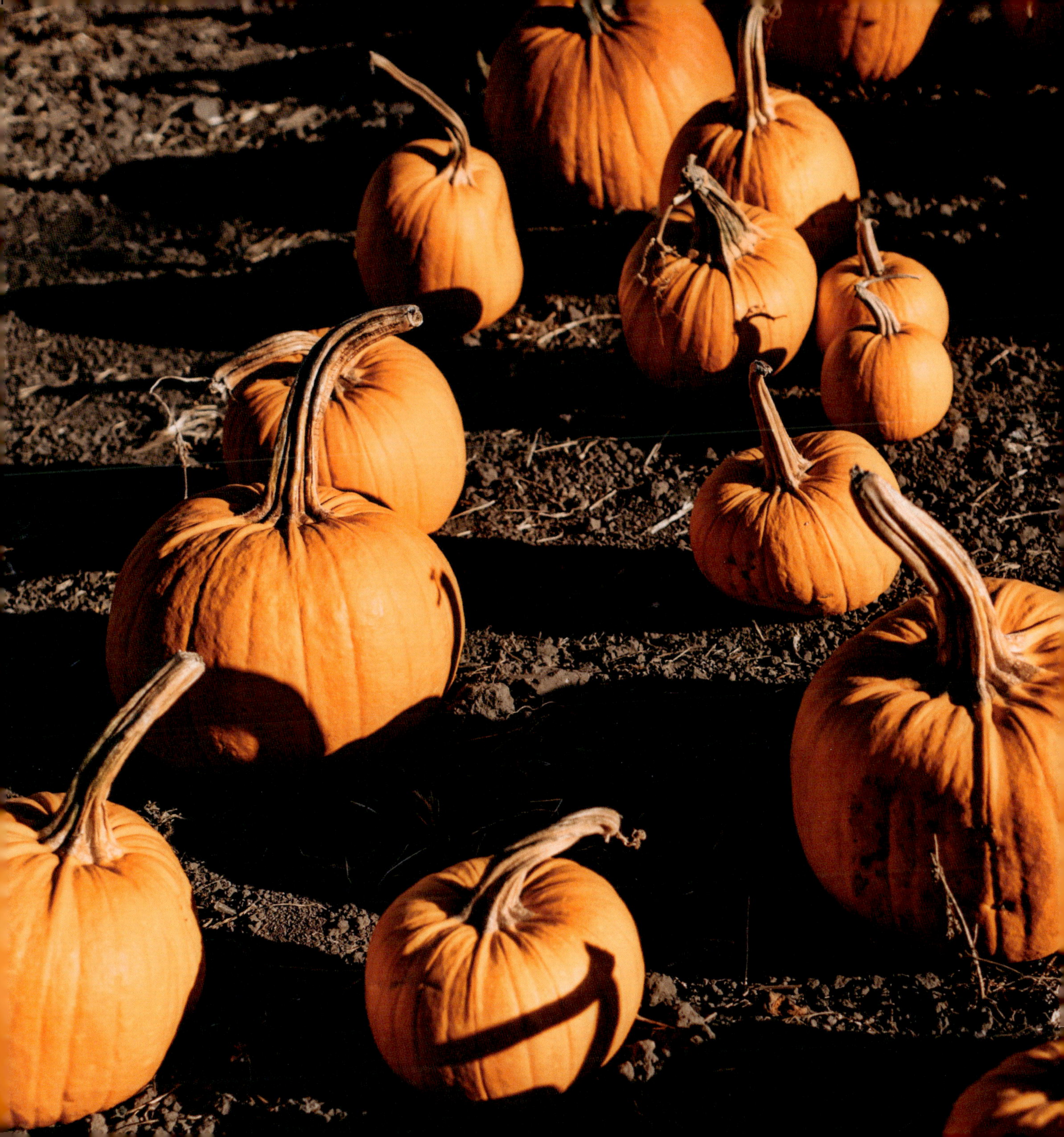

I **see** calico corn!

I **think** the kernels are different colours.

I **wonder** if there are more purple than orange.

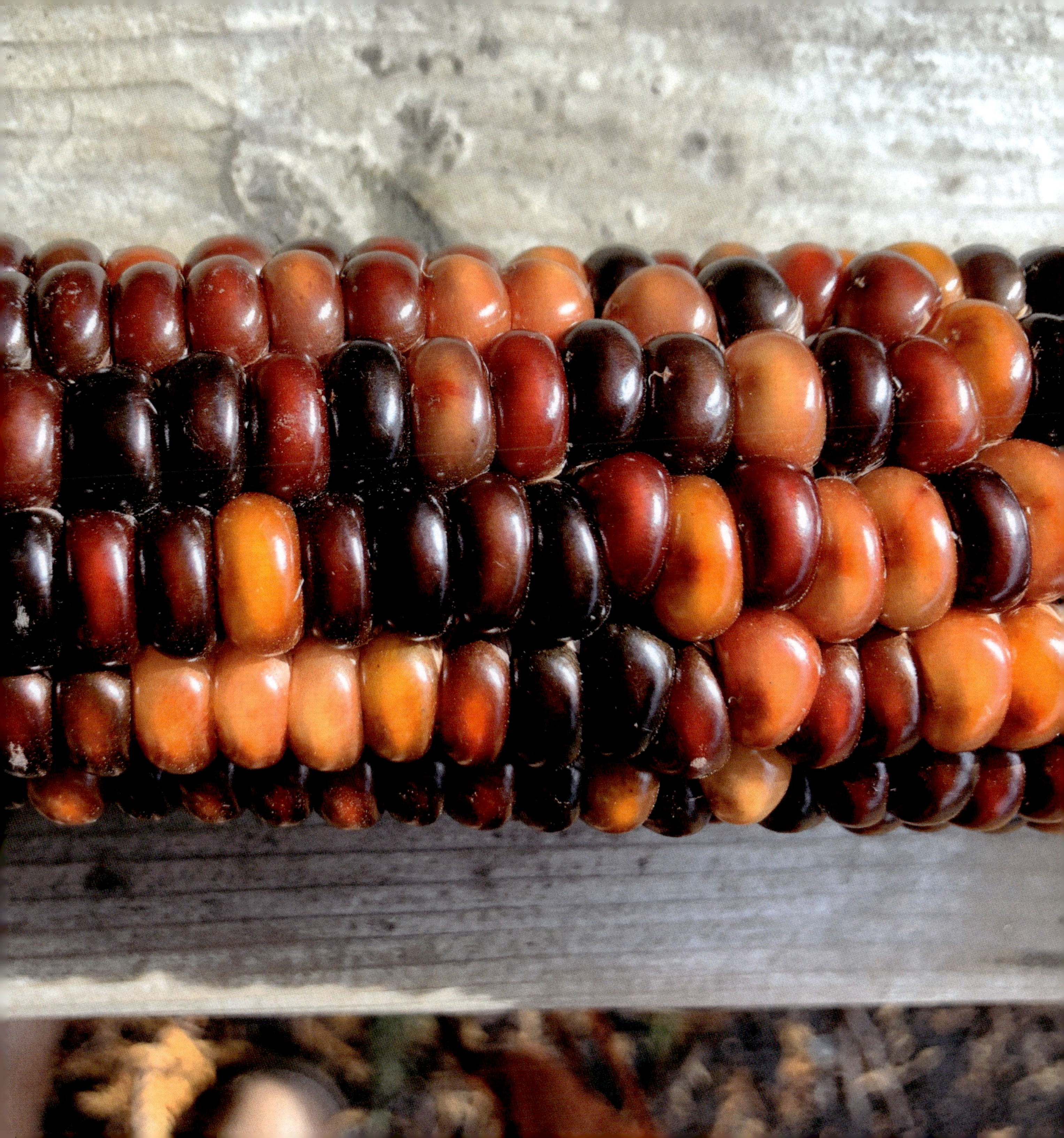

I **see** a yellow gourd!

I **think** that it has a long stem.

I **wonder** why it grows in a patterned twist.

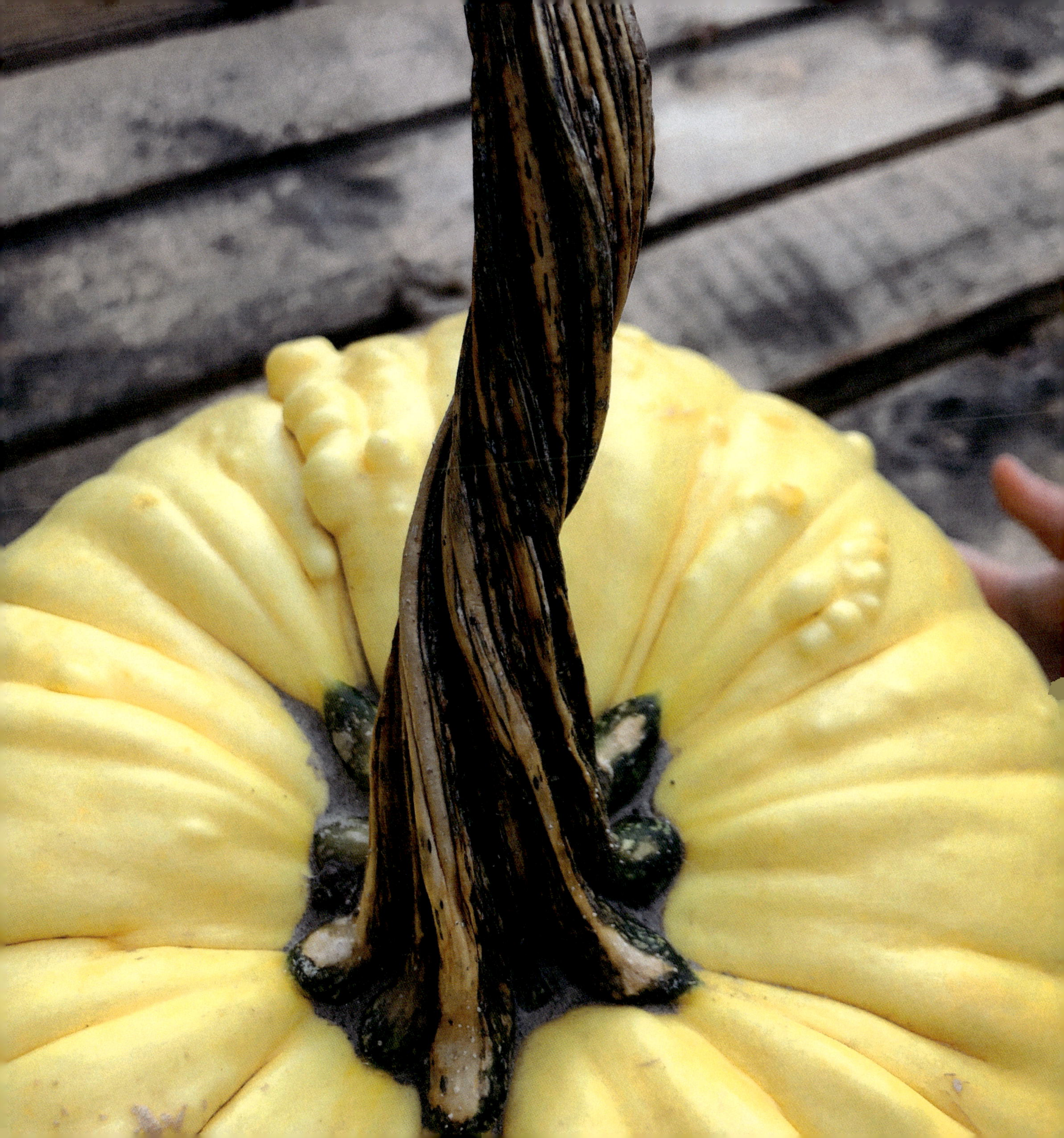

I **see** a flock of birds!

I **think** they are travelling together.

I **wonder** how they know which way to go.

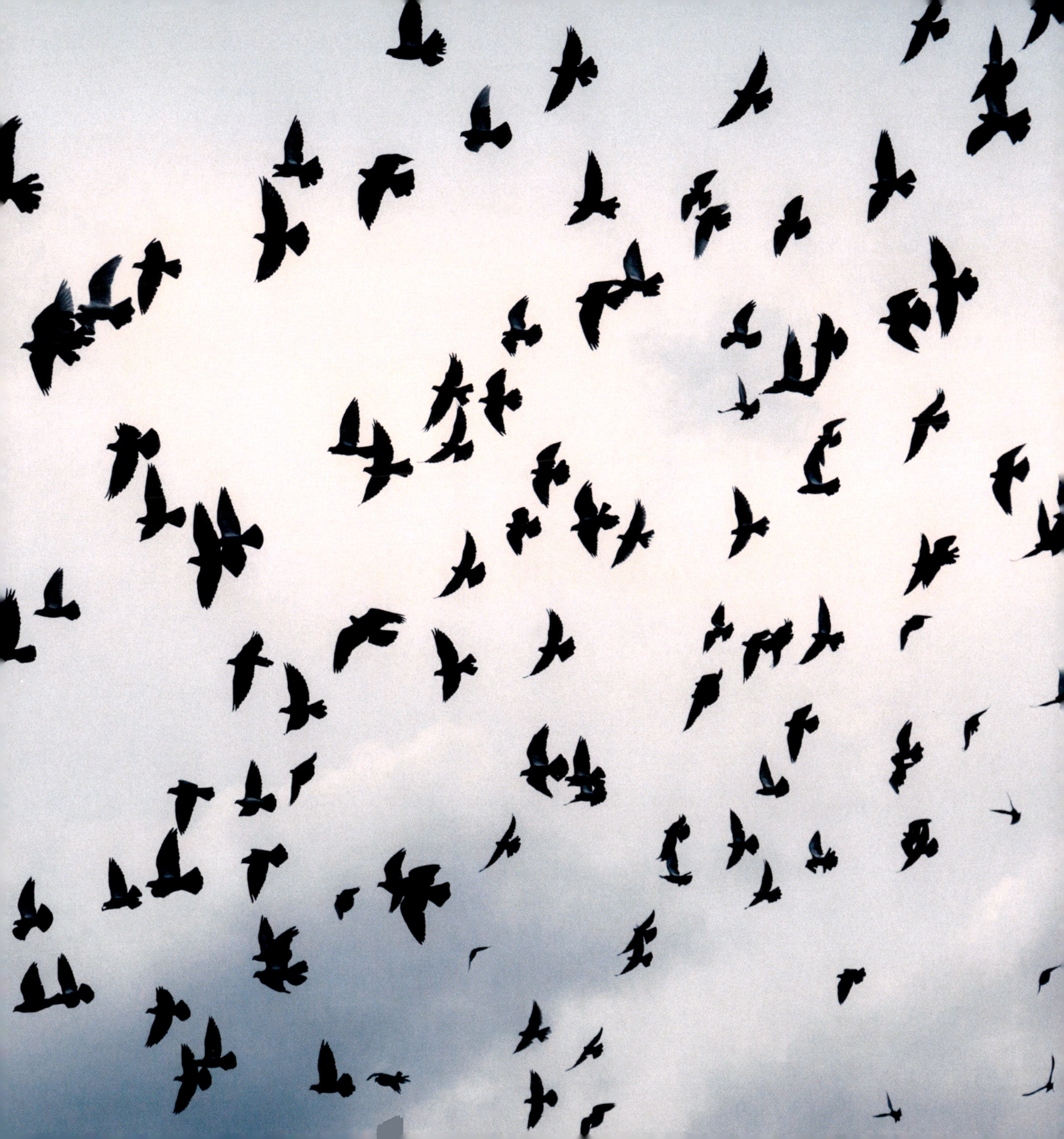

I **see** a spider's web!

I **think** it is covered in dew.

I **wonder** how many little droplets it is holding.

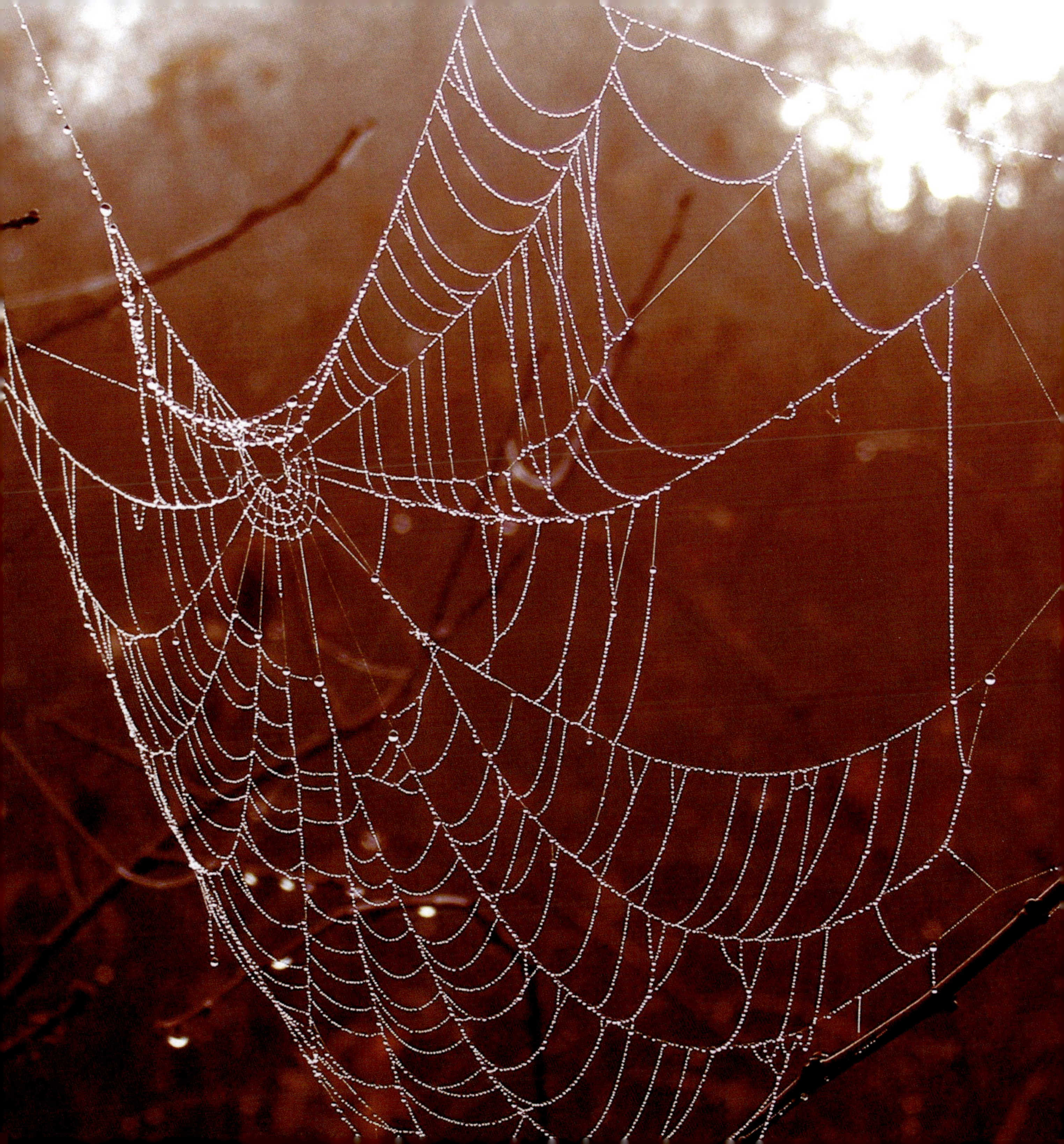

I **see** a brown spider!

I **think** it has eight legs.

I **wonder** how high and fast it can jump.

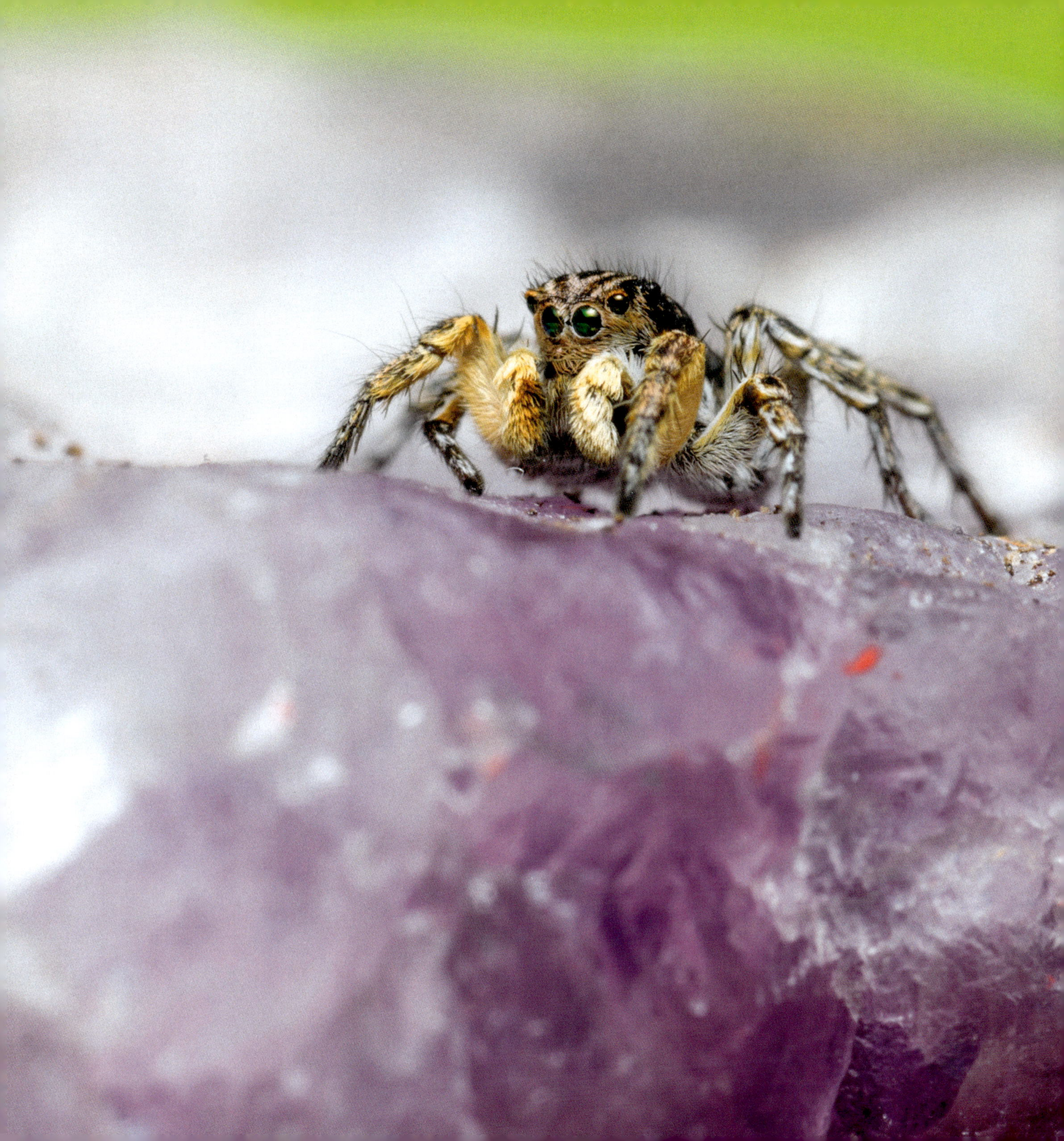

I **see** an orange pumpkin!

I **think** it is hollow inside.

I **wonder** what shapes I can carve into a face.

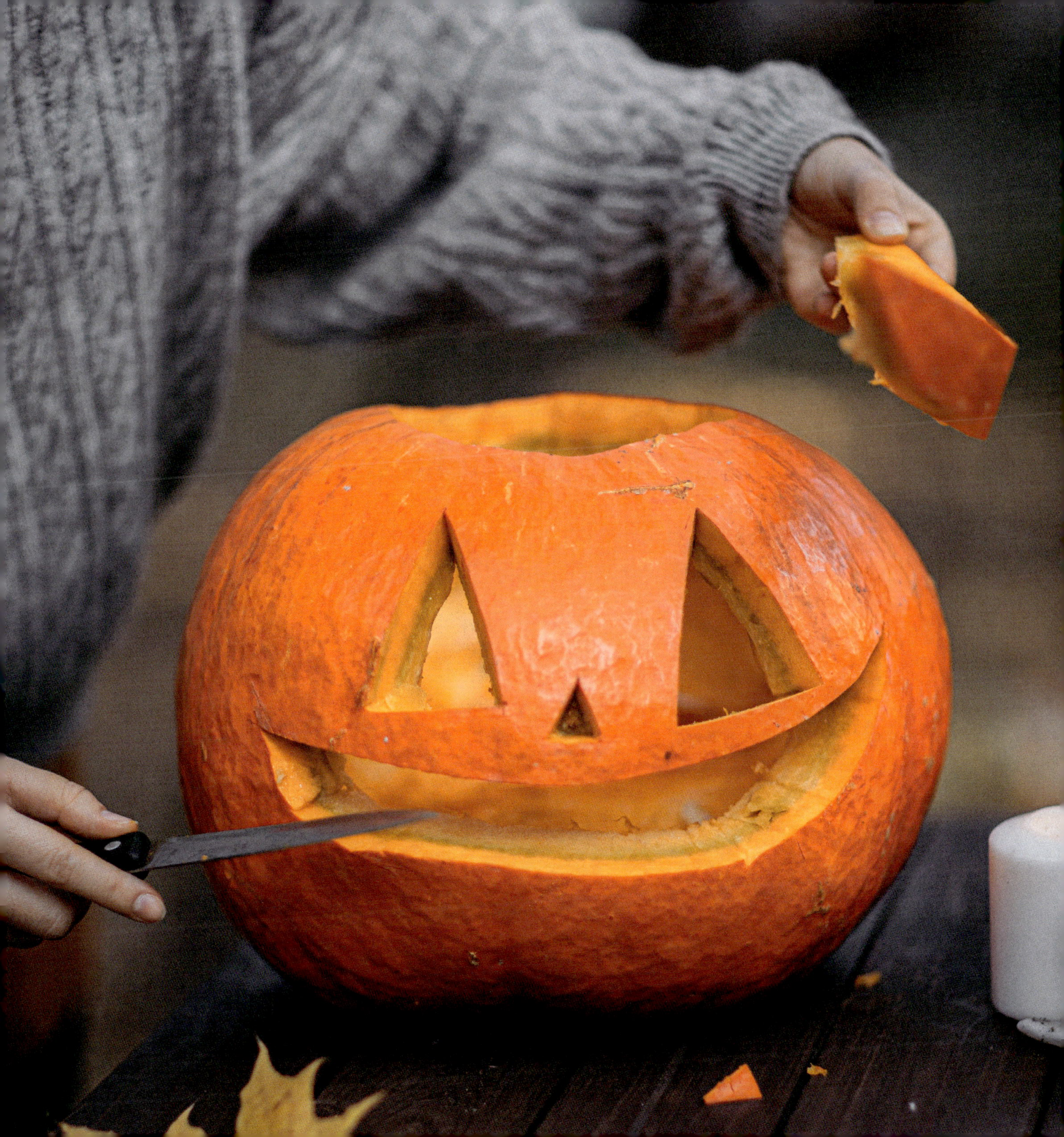

I **see** a jack-o-lantern!

I **think** there is a light inside.

I **wonder** how brightly it glows.

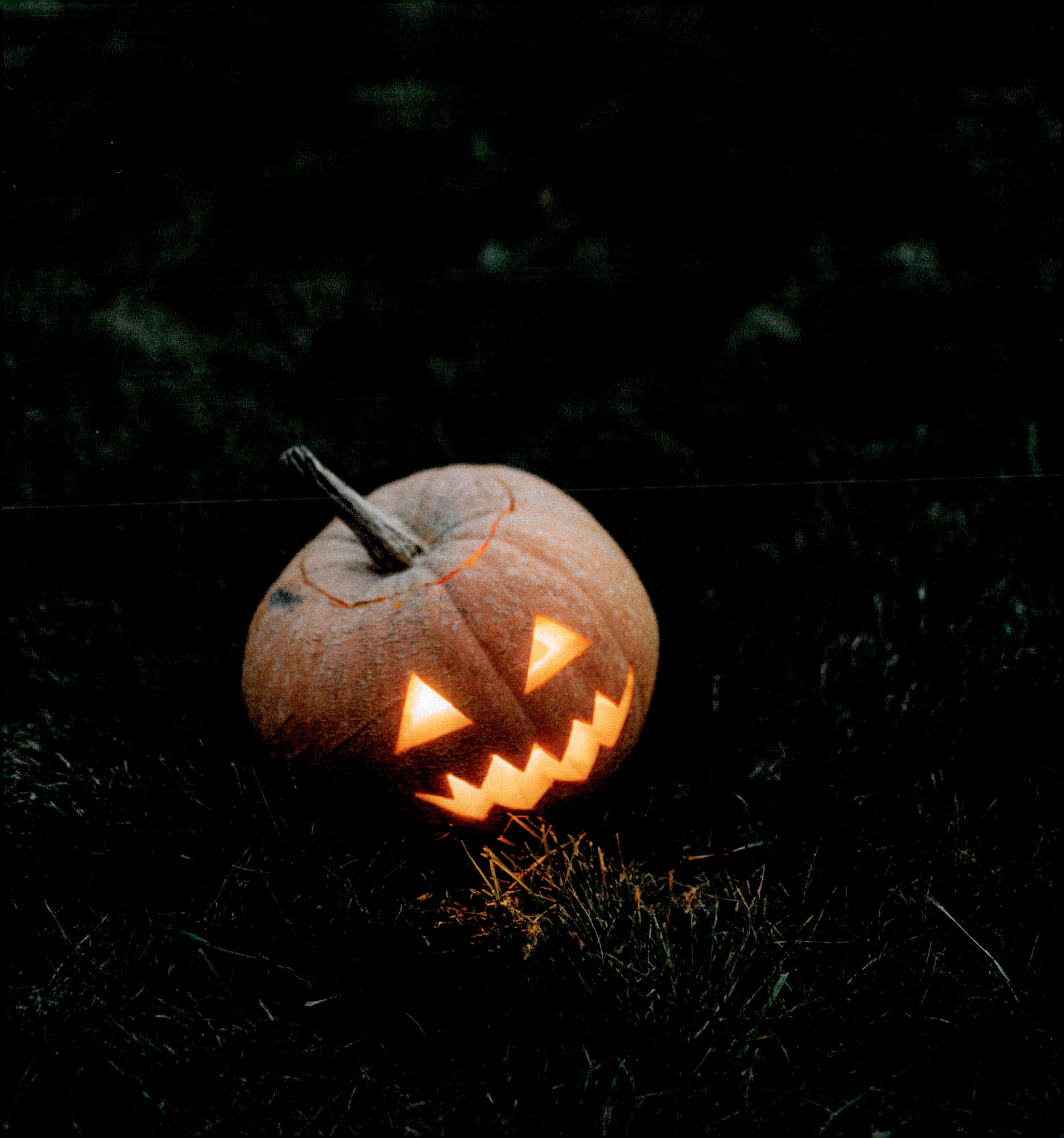

I **see** trick or treaters!

I **think** we are walking all around town.

I **wonder** how much candy we will collect.

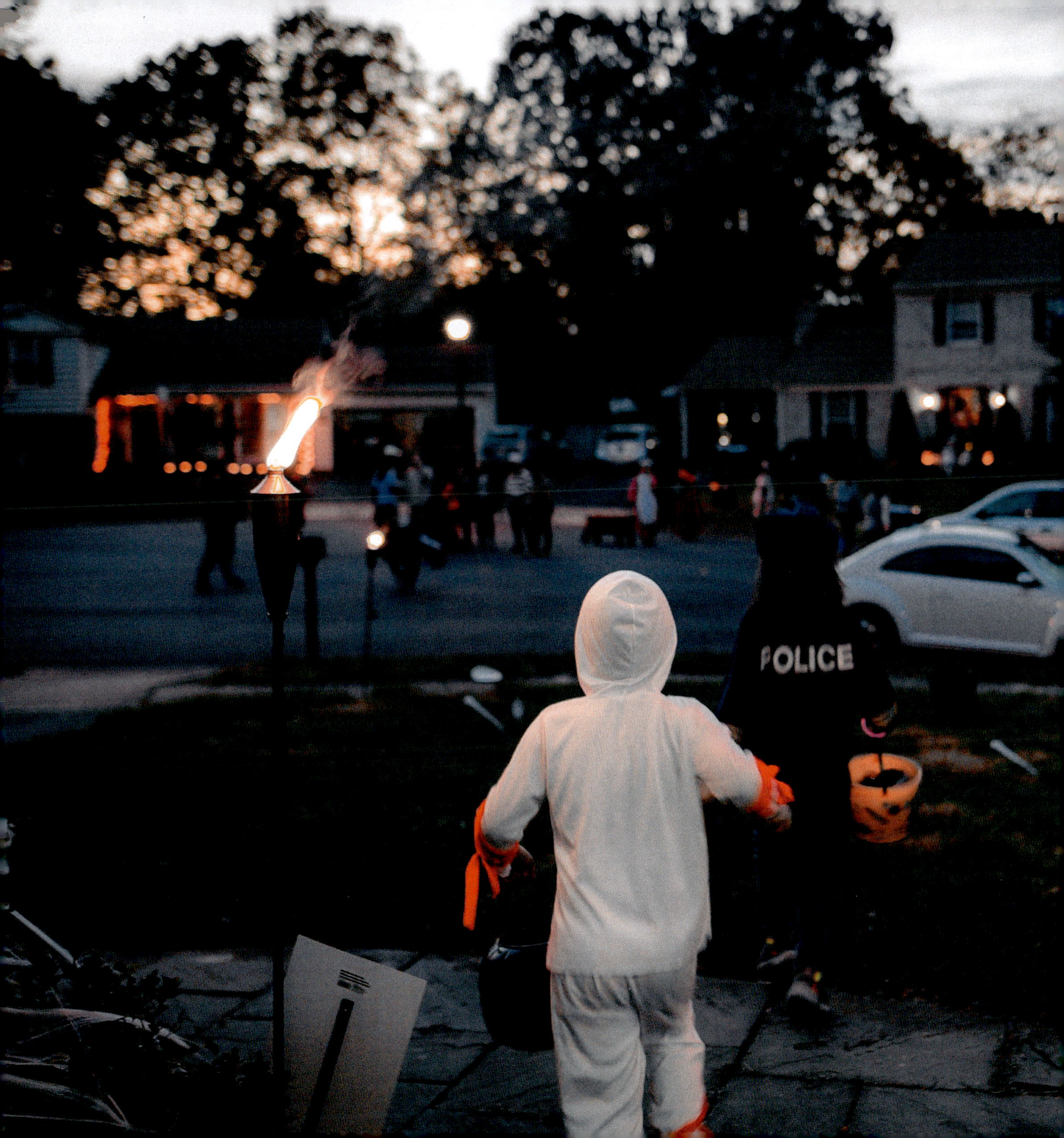
POLICE

I **see** candy in my bag!

I **think** there are many different kinds.

I **wonder** how long until I eat it all!

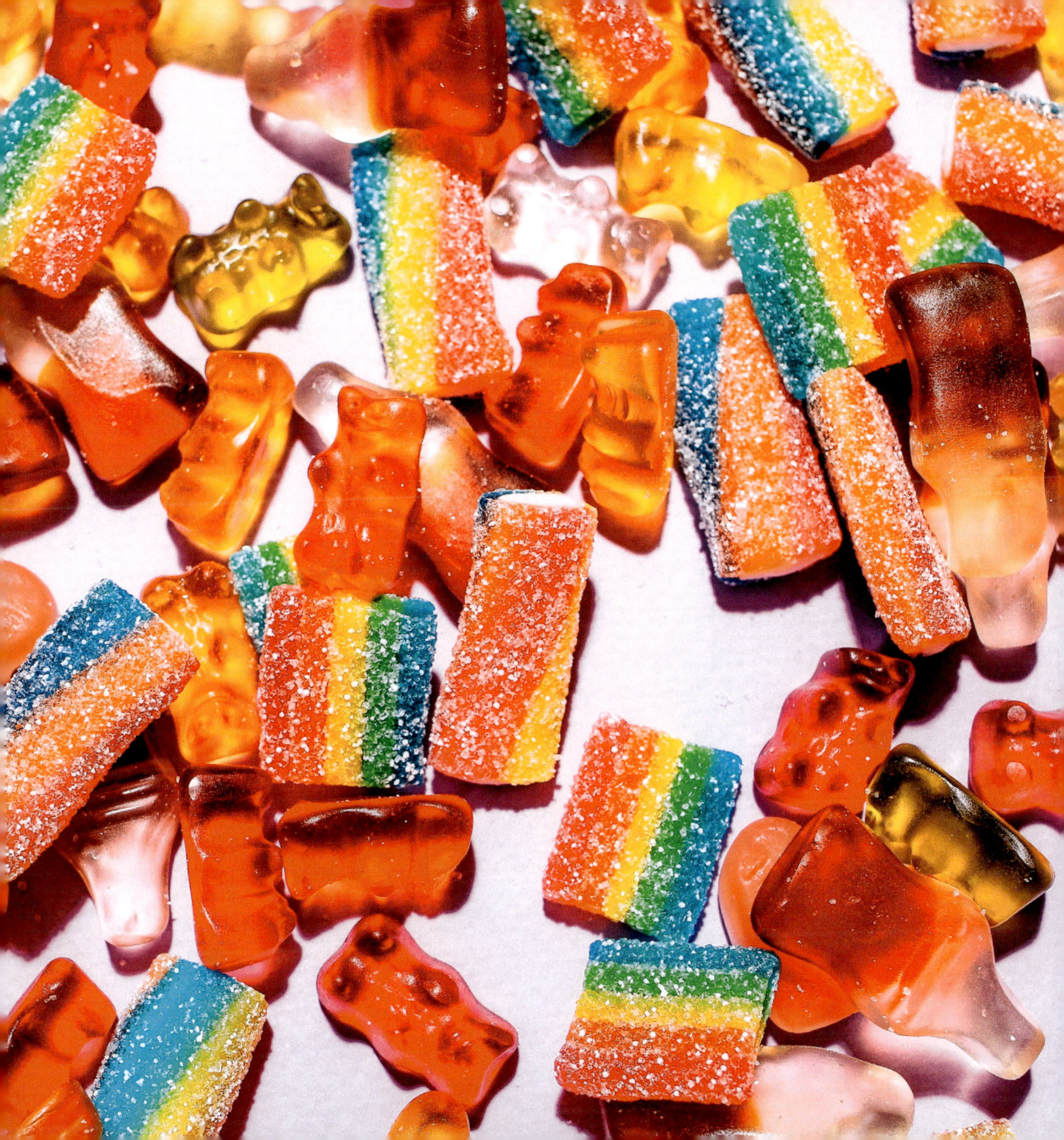

Author's Note

Math is all around us! As an educator I love helping children discover the authentic ways we use math in our everyday lives! As children recognize the integrated, meaningful ways math helps our world work, their interest and confidence in the subject will grow. Exploring the authentic math that exists in our surroundings may help nurture children's interest and confidence, building a strong foundation for subsequent experiences. I love taking math out of 'math class'.

My hope in writing this book is to inspire children, educators and families to see math as an inviting discipline that does not exist in isolation. Seasonal experiences including Halloween offer countless opportunities for children and adults to make amazing mathematical connections.

This book does not need to be read beginning to end. The photos can be used individually, or in combination, to spark mathematical conversations and connections with children. Ask children what they **see, think and wonder** about each picture. Ask what their theories are for what they see happening on each page. Adults can support and extend children's mathematical and scientific ideas by asking them to share their observations using the see, think, wonder routine.

The information presented in this book can serve as an introduction to new math concepts, or as a reference when mathematical concepts are discovered by children as they explore Halloween pictures, experiences and activities (e.g., a visit to the pumpkin patch, a costume parade). You might choose to use only the photos as conversation starters one at a time, or read the book in its entirety using all photos and narratives. Perhaps children will be inspired to create their own 'Halloween Math' book unique to their specific context. The possibilities are endless!

When we look at the world through a mathematical lens, we discover that anything is possible!

-Deanna

How to Use This Book

This book can be read to children using the '**see, think, and wonder**' sentence starters that correspond to each picture. Children can be asked to consider the mathematical prompts on each page, and hypothesize about how they might research the inquiry that is presented.

This book can also be shared with children using only the photos. Present each photo to children one at a time. Engage children in a mathematical conversation using the see, think, and wonder routine as they explore the photos.

At first children can be invited to carefully observe each photo and share what they **see**. Ask children to use rich description as they articulate their observations. Next, ask children to make personal connections to the information presented in the text and photos. They can articulate what they **think** about the question prompts in the text, or make inferences about the information shared in the photos. Finally, ask children to share what they **wonder** about the text and photos.

What are children curious about? What do they notice in the foreground, and background of each photo? What connections can they make to the book? What experiences do they have that relate to the objects or situations being presented? Is there something they are interested in learning further? How might they go about conducting mathematical research? What knowledge do they need to have in order to research their question? What tools and supports might help them in their quest? How can they share their findings with others?

This rich mathematical conversation can inspire children to understand what they read, make connections to the book, and inspire inquiry-based learning for deeper exploration and understanding.

After the children have explored the book, consider asking them to co-create their own version of the text. Children can research additional Halloween artifacts and experiences, illustrate pictures and write their own narratives. Invite children to look around their homes and communities for other seasonal situations to explore. Perhaps children can digitally document what they find and add these to their own Halloween Math book. The possibilities are as endless as the questions children ask.

Deanna Pecaski McLennan, Ph.D., is an elementary educator in Ontario, Canada. Deanna is fascinated by math and loves exploring its natural and authentic application in the living world. She hopes to help children and families recognize math as a beautiful and fascinating subject, and grow children's confidence, accuracy and interest in math.

Follow Deanna on Twitter and Instagram for regular updates including ideas for engaging children in playful, emergent math inside the classroom and beyond. Extending math learning outdoors is a favourite exploration!

Connect with Deanna:

deannapecaskimclennan@gmail.com
@McLennan1977

Also from Deanna

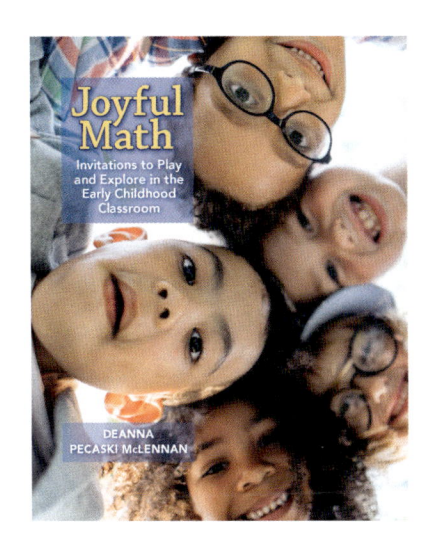

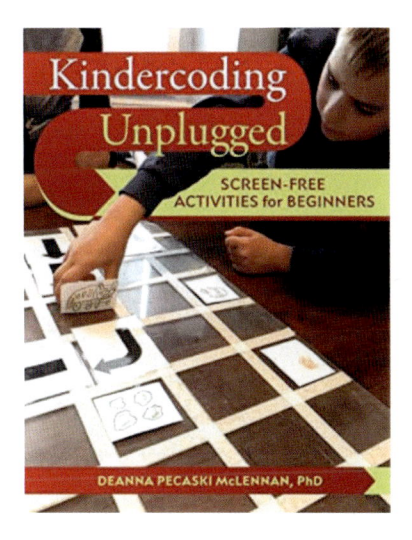

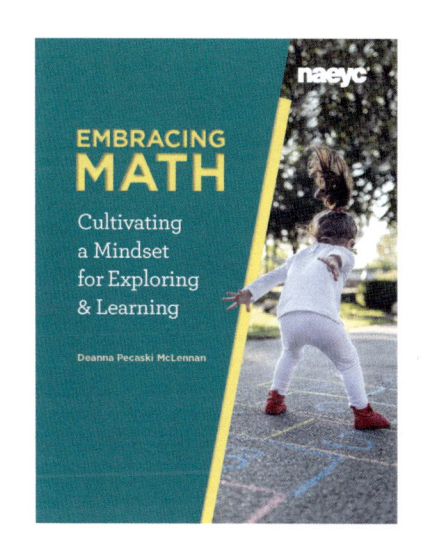

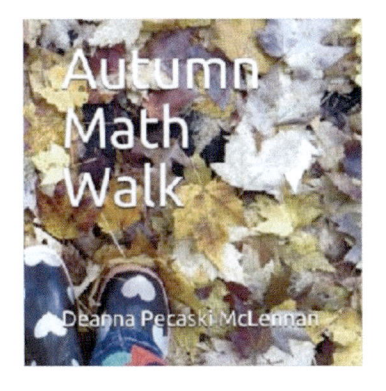

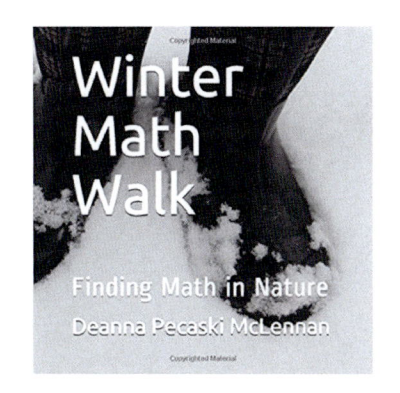

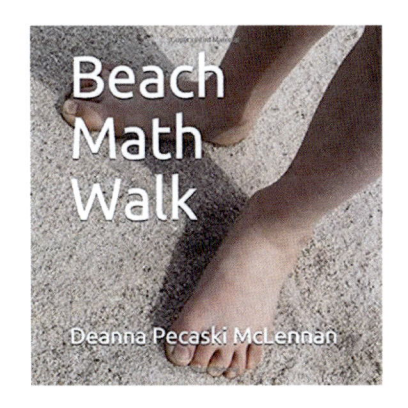

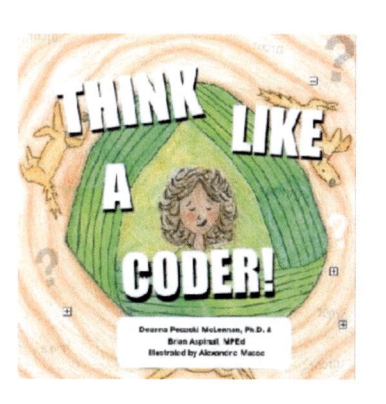

Joyful
Math

Manufactured by Amazon.ca
Bolton, ON